W9-BFQ-691

HOW TO BUILD
CLOCKS AND WATCHES

HOW TO BUILD CLOCKS AND WATCHES

by
BYRON G. WELS

A VERTEX BOOK
Princeton New York Philadelphia London

Library of Congress Card Number: 73-124636
International Standard Book Number: 0-87769-038-3

First printing

Printed in the United States of America

to the memory of Norman Rubin

*A man's worth is not measured by
the number of his years on earth.*

Contents

Foreword

THE AUTHOR wishes to express his thanks to the following companies, which cooperated in providing most of the materials for this book. The reader wishing to duplicate any of these projects would do well to start by writing to these firms for their catalogs, from which all the parts used were drawn.

- Edmund Scientific Company
  100 Edscorp Building
  Barrington, N.J. 08007

- Lafayette Radio Electronics
  111 Jericho Turnpike
  Syosset, N.Y. 11791

- Lanshire Clock & Instrument Co.
  c/o Empire
  1295 Rice Street
  St. Paul, Minn. 55117

HOW TO BUILD CLOCKS AND WATCHES

Introduction

MAN HAS ALWAYS been fascinated with the passage of time, for it indeed marks his duration on earth. To mark the passage of time, he has developed some rather ingenious devices. One device employed water dripping at a given rate into a container. The container walls or the walls of the supply vessel were calibrated in time divisions.

Certainly, we are all familiar with the hourglass and its dripping sands. Candles have been marked off to tell time, and indeed, as oil will burn at a regular rate, so have oil lamps been used as clocks.

Man has also marked his time with the passage of the moon, as did the Indian inhabitants who originally settled this country. The sun was more popularly used in Europe, where the sundial was developed.

And then came the escapement. With it, man could store energy in a large spring and pass this energy into a gearlike wheel, called an escapement. A pawl, or rocker, operated in this wheel's teeth, allowing the wheel to rotate a small amount with each movement of the rocker. Additional gears allowed this energy to be transmitted to hands, and the concept of the modern clock was born. The familiar "tick-tock" is the sound of the pawl operating in the escapement wheel.

But man was not satisfied. He harnessed this stored energy in other ways as well. Large, beautiful clocks were developed, clocks which had mechanical manikins to strike the hours, or small figures to move about the clock as it operated. In Germany, the time-keeping function of the clock was all but disregarded, as wood-carvers strove to outdo each other in using clockwork mechanisms to animate their products.

Time, which is essentially what we're talking about, has seen a great many changes take place. With electricity, electric clocks became the fashion, and were soon used in homes and automobiles. The electric clock used in the home is very different from the electric clock used in an automobile. The electric clock used in the home has a synchronous motor which draws its accuracy from the never-fluctuating 60-Hertz frequency of alternating current—and no escapement wheel is required. As there is no 60-Hertz electricity available in a car, which is equipped with a direct-current battery, an electric wind-up system is employed. An electric motor winds the clock's spring until sufficient tension is applied to keep the clock going. When the clock runs down, the motor starts again and rewinds the spring.

Man is still looking ahead; transistorized clock movements are with us today, operating for as long as a year on a single energy cell. Electric wristwatches are operated the same way.

Aware that certain things are unalterable in their time-keeping course, man has been able to produce the cesium clock, which depends on the rate of decay of radioactive elements for accuracy. A modern watch using a tuning fork instead of an escapement wheel is also currently popular.

We have combined clocks with mechanical devices to sound alarms to wake us and with radios and television receivers, and the outlook is still progressing toward an unforeseeable end. We can, at this point, only speculate.

HOW TO USE THIS BOOK

In this book, you will find what we believe to be some pretty unusual clocks. More important, these are clocks that you can build for your own home, as all the necessary construction information is given with each project. Diagrams and step-by-step photos make everything very clear.

The heart of any clock is the clockwork mechanism, and happily you can buy a working mechanism ready to install. At the back of the book are complete lists telling you where to buy the necessary components for the clocks you wish to assemble. Rest assured that not only will these clocks be attractive and accurate, but they'll start people talking as well.

Start by going through the book and reading at least until you understand how the particular clock in which you are interested operates. Then, should you decide to build one of these units, check the bill of materials, and get started with the instructions. You'll find that all the parts are readily available.

WHAT KIND OF MECHANISMS?

There are three basic types of mechanisms available to you. The simple electric type which must be plugged into a wall outlet, the transistorized type that operates from a self-contained battery source, and the spring wind-up type. Consider the type of clock you are going to build, and select a clock mechanism accordingly. If you feel that a line cord connecting from the clock to the outlet would be an eyesore, then choose the battery-operated type. If price is a problem, the wind-up type may be substituted. In any case, you should have no trouble at all in using an old clock that you have around the house, provided that it still operates.

SOME WORDS OF ADVICE

Frequently, an electric clock is discarded because it starts making unwelcome noises. If these noises are the result of the gears gnashing their teeth, chances are that you were right to retire the clock in the first place, and you would be ill-advised to build it into a new clock. However, some clocks only need to be lubricated, and this can best be accomplished by allowing the clock to run upside-down for a while so that the oil, following the tug of gravity, will again flow to where it belongs.

Where possible, in all these projects, build with a good deal of care. Neatness certainly does count, for the care taken in workmanship will be reflected in the finished product. If you exercise sufficient care, you can puff up your chest while people "ooh" and "ahh," and not have to spend valuable time in making excuses for unwarranted scratches or other goofs!

A WORD OF CAUTION

While all the projects in this book have been well tested an element of caution must apply. In working with any tools, be careful and use the good judgment that any craftsman must apply. In most of the electrical circuits used in this book, you will be dealing with 117 volts A.C. Be careful not to work on "hot" circuits; and where switches are used to control the A.C. line voltage, do not rely on the switch alone to cut off the power, but remove the plug from the outlet as well. After all, unless properly polarized, the wall plug could be installed in a reverse fashion, putting the switch in the cold side of the line. Inadvertently touching the wrong part of the circuit might make *you* the shortest path to ground!

LET'S GET STARTED

While the author has endeavored to explain each technical term the first time it is used, refer to the glossary at the back of the book for further clarification if you need it.

And if you run into any specific problems in building your clocks, do not hesitate to contact the author by writing to the publisher; you will receive an answer by return mail. And now, let's go!

PROJECT I

The Picture Clock

THE PICTURE CLOCK is a combination of a clock and a photograph that you can hang on your wall. People who look at it will think it is an interesting photo, but it's more than just a photo, for with it, you can tell the time!

To start with, you should have a photograph of a well-known clock in your home town. The author selected the clock that sits on top of the information booth at Grand Central Station in New York City. Take your photo using a relatively good camera with a sharp focus and use a relatively fine-grain film, such as Plus-X, so that you can have the picture enlarged suitably. Make sure that you get in close enough so the clock in the picture will, when the enlargement is made, be at least two inches in diameter. Shoot the clock head-on so that it will be a perfect circle, not distorted into an ellipse.

INFORMATION
INFORMATION
INFORMATION
INFORMATION

The photo can be made in black and white or in color. When you select the best negative, take it to your local photo-processing shop, and have the picture enlarged to at least sixteen by twenty inches.

When you get the picture enlarged, use rubber cement, and glue the photo, back down, on tempered hardboard. Cut the hardboard so that it is flush with the edge of the picture. To make the picture a bit larger and to provide an interesting matte border, use textured Marlite paneling, available from your local lumber dealer.

When the cement has thoroughly dried, drill a small hole through the face of the clock in your photo, and using a fine-toothed coping saw, cut out the entire clock face, following the circle of the clock to make the hole as round as possible. Now insert the clock you select from the rear of the board. Choose a clock whose dimensions are as close as possible to the circumference of the hole. If you have a two-inch hole, choose a two-inch-diameter clock. Mount the clock from behind, and attach the bezel, or metal ring and glass, from the front. If you have cut accurately and chosen a clock with care, the electric clock will actually look like part of the photo. The illusion will be improved if no electric cord shows. On one model, the author had a wall outlet mounted at the clock's level, rather than at the baseboard, with the result that no electric cord was visible. Another possibility is to substitute a transistorized (battery-operated) movement.

To complete this project, obtain metal framing sections from your photo or hardware dealer, and frame the board with this material trimmed to size. Hang the picture on your wall, set the clock, and plug it in.

Then invite some friends over. It's compliment time!

MATERIALS

1 photograph suitably enlarged, in black and white or color, in which a clock is prominently displayed
1 piece of tempered hardboard, such as Masonite or Marlite paneling, of a suitable size
1 clock mechanism, complete with glass face, bezel, and mounts
1 picture frame

PROJECT 2

The Einstein Clock

ALBERT EINSTEIN, the noted physicist, could not be bothered with petty details, and he made the point that a clock with one hand was quite enough for him.

It wasn't that he was unconcerned about the accurate time, but he had developed a tongue-in-cheek solution to the problem that could easily be explained by physics —and his one-handed clock was able to provide him with as much accuracy as two-handed timepieces did for others!

The Einstein clock is very large. It is this largeness that provides the accuracy. Consider that the divisions between the minute symbols on a clock or a watch are nothing more than degrees of a circle referred to the center of that circle. The closer you get to the center of a circle, the closer together those divisions become. The further away, the more distance there is between them.

In an accurate timepiece, when the minute hand says "half-past the hour," the hour hand is exactly midway between the hours indicated.

The Einstein Clock has only an hour hand and a large clock face. The clock face is divided in such a way that there are sixty minute divisions between each hour division, with the five-minute segments accented with heavier lines.

The large, single hour hand moves slowly around the clock and is accurate to the very minute should you require this exact information. Yet the Einstein Clock is just as easy to see for determining the approximate time at a glance.

To build this clock, you will need a fairly powerful clock motor to drive the heavier hour hand that you will be making.

Visit your local electrical supply shop and obtain a wall-mounting box of the type used for mounting a cluster of four receptacles. This should prove sufficiently deep to properly house the clockwork mechanism.

Locate a stud in the wall by tapping the wall with a mallet until you hear a deeper thud. Cut through the wall at this point and mount the box against the stud so that it is rigidly supported. Use BX or Romex cable (check with your local wiring codes) to provide power inside the box from a nearby outlet.

Obtain a large metal cover plate that will attach to the face of this box and will be of sufficient size to cover any marring of the wall caused during the last operation. This plate is also available at most electrical supply houses and is usually painted with a prime coat, ready for finish painting.

Drill out a one-quarter-inch hole in the center of this plate and mount the clock so that the armature which supports the hands will protrude through the front. Attach the clockwork mechanism to the back of the plate, and wire it to the cable you just installed. With the power applied, the clock motor should run. (Note: If you use a transistor-powered motor, you won't need the A.C. wiring!)

The next step is to prepare the clock face.

Tie a thread about four feet long to the center armatures of the clock, and then tie a pencil to the other of the thread. By keeping the thread taut, you will be able to scribe a perfect circle around the armature. Wind some of the thread around the pencil so that the thread is about two inches shorter, and repeat this process. You will have two concentric circles, two inches apart.

Using black India ink and a Speedball pen, carefully go over these lines until both circles are drawn in black ink.

Drop a vertical line from the top, through the armatures to the bottom, and mark the position for the numerals "12" and "6" in pencil. A horizontal line through the middle will give you the "9" and "3." To get the other increments, a 30° separation with a protractor can be used, and a bit of head-knocking can even provide you with the other increments. For most practical purposes, dividing the hour spaces into quarters is usually sufficient.

When these have been marked with the India ink, you can use the ink and the Speedball pen to identify them, or, if you are not of an artistic bent, a visit to your local art supply shop will get you a couple of sheets of rub-off type with some fancy Roman numerals. Follow the instructions and put the numbers into place.

Part of the accuracy of this clock depends on the fine line of the hour hand used. Select a piece of wire painted black. To get the wire dead straight, tie one end to a door knob and the other end to the shank of a screwdriver. Then pull. The little kinks and bends all disappear, and the wire straightens out beautifully and holds its shape. Cut the wire from the knob and from the screwdriver, and try to keep it straight.

Now place the wire parallel to the hour hand from the clock, and either solder or cement it to the hour hand. A small rubber band will hold it in position while it dries.

The hour hand is held to the armature by friction, and all you have to do is slip it into place.

For the greatest accuracy, wait until the time is exactly on the hour, and set the hour hand to that time. Make all adjustments in position by grasping the hour hand near the hub to avoid bending the wire.

This clock was built by some of our friends who are married, and the consensus is that you had better build this one while your wife is out. Women seem to object when you begin to hack at the wall, but be fearless! The finished job is worth it.

Want proof?

Invite some friends over—it's compliment time again!

MATERIALS

1 clock mechanism with hour hand only
1 wall-mounting electrical box, square
1 large cover plate for above box
1 bottle of black India ink
1 set of rub-off numerals, Roman or other
1 Speedball pen
6 feet of solid electrical wire

PROJECT 3

The Night Watch

THE CHANCES ARE that you have a clock or watch with a luminous dial. If you expose that dial to light and then look at it in the dark, it will glow (more or less) for awhile. But try—just *try* to wake up in the night and see the time! If your clock or watch is like most available today, the rate of decay of the luminance is so rapid that you won't be able to make out the time.

This project eliminates the problem entirely and uses your own clock or watch. The heart of the system is an electronic flash unit. The one we used is available at a cost of only $9.95. This is mounted in a small box that will look all right on your night table. A small cigar box covered with adhesive-backed plastic will do nicely. Drill a small hole, one-quarter inch in diameter for the switch. Mount the switch in the box as shown, and connect the switch contacts to a short length of electrical "zip" cord.

Put a standard A.C. plug at the other end of these wires.

Place the flash unit in the box so that it faces the front and mount it tightly in place with Pliobond cement. Now cut a one-inch hole in the top of the box. To complete the job, cement a small mirror to the front of the box, allowing the wires to pass around the mirror's side.

Close the box, and mount the assembly as shown.

To use the unit, place your bedside clock on top of the box, in front of the bracket. If you prefer to use your wristwatch, place the watch over the bracket so that the light from the flash unit will bounce off the mirror and reflect to the watch.

Now it's night time, and you're tossing and turning. You want to know the time, but can't make out the hour on your luminous dial. Close your eyes so the flash won't blind you, or turn your head away. Reach out and press the button. The flash goes off, the light bounces from the mirror up to the clock or watch, instantaneously recharging the luminous dial, which you can now read with ease.

The flash unit will give thousands of flashes before it needs replacing, and the unit itself, which operates off batteries, will normally need a change of batteries about once a year. For even longer life, you might wire another on-off switch in series with the switch on the flash unit, and mount this switch on the box so you can turn it off during the daylight hours when it is not needed.

Incidentally, the flash is of such short duration that the chances are that it will not disturb others who sleep in the room.

MATERIALS

1 photographic electronic flash unit
1 small box
1 foot of electrical "zip" cord
1 A.C. plug
1 momentary-contact SPST switch
1 bottle of Pliobond cement
2 square feet of adhesive-backed, decorated plastic (Contac or equivalent)
1 small mirror
1 6-foot A.C. cord and plug set (comes with flash unit)
1 small wooden bracket for wrist watches, made from ½-inch dowel

PROJECT 4

The Ceiling Clock

THE CEILING CLOCK uses an inexpensive opaque projector. The projector we used operates with a single 60-watt lamp and has a built-in slide at the base. From black cardboard of the type that comes with photographic enlarging paper, cut a piece that fits easily into this slot.

Before finally assembling the unit, mount it on the floor near your bed so that the lens points upward. Turn the switch to the ON position, and then unplug the electric cord. The lamp will go out.

Carefully cut *one side* of the line cord, and strip the ends of this wire. Attach another length of wire to these ends, and carefully tape the connections. Attach a pendant-type push-button switch to the remaining ends of this extension, and place it convenient to the bed. To test the unit, plug it into a wall outlet. When the pendant switch is depressed, the projector should light.

To use the unit, place your wristwatch over the black cardboard slide before retiring at night. Arrange it so that the face of the watch faces the projector's lamp. Should you wish to see the time at night, simply press the pendant switch, and you can read the exact time on your bedroom ceiling!

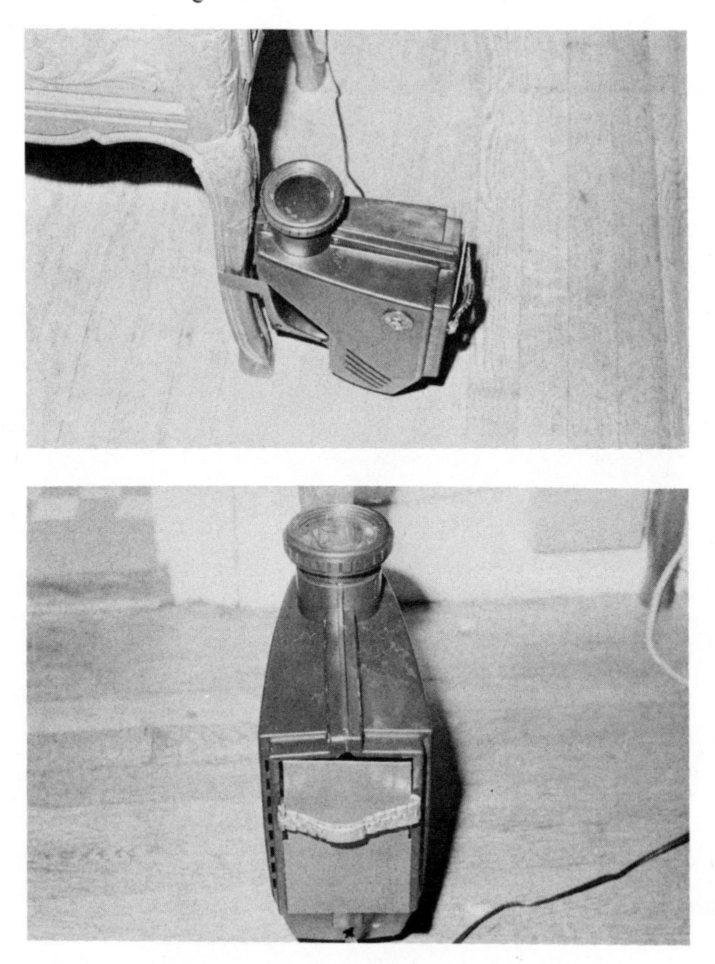

MATERIALS

1 opaque projector
1 momentary-contact SPST switch

PROJECT 5

The Electric Sundial

CAN YOU IMAGINE an electric sundial? Probably this novelty clock is the ultimate in modernization. What's more, it really works, and it keeps time accurately. The trick is that instead of the light source (usually the sun) moving around the fixed dial so the pointer shadow moves to tell the time, this electric sundial rotates. The light source (a bulb) and the pointer are fixed!

To build this clock, start with a suitable base. You can easily use a fairly deep cigar box for a base, and either paint it or cover it with a suitable vinyl plastic with an adhesive back.

Cut out the top of the box and mount the electric clock mechanism so that it protrudes. Using a simple pair of cable clamps, attach a short length of flexible gooseneck to the back of the box. Mount a light socket with a switch to the top of this, and pass the electrical cord through the gooseneck. Run this cord through a small hole in the back of the box, and connect it in parallel with the terminals of the clock motor, then back out to a line cord and plug. As you can see from the circuit diagram, the clock motor will operate all the time as long as the cord is plugged into a wall outlet, and the lamp can be turned on or off at will.

Obtain a lightweight but rigid disc, which will serve as the dial. Locate and drill the center, taking care that when the disc is mounted on the armature of the clock motor, it will clear the lamp gooseneck.

Using the Pliobond cement, attach the disc to the top of the clock's hour hand, centering it carefully. Allow the cement to dry overnight at least.

Now you can decorate the disc in the manner of typical sundials. You might start with a base coat of gold enamel and use black paint to apply the numerals and curlicues. When the paint has dried, press the disc in place on the hour hub of the armature.

The pointer can be made of any wood or metal scraps, but keep it thin for more accuracy, and give it the same decorating scheme you used for the dial.

To use the clock, either set it on a table or hang it on a wall. Plug it in, and rotate the disc so that when the light is turned on, the shadow of the pointer falls across the face of the disc and indicates the correct time,

For a broader shadow, try offsetting the lamp a bit. The unit makes an interesting conversation piece and is an accurate timepiece as well. You can turn the lamp on or off, but of course, to get a reading of the time, the lamp must be on.

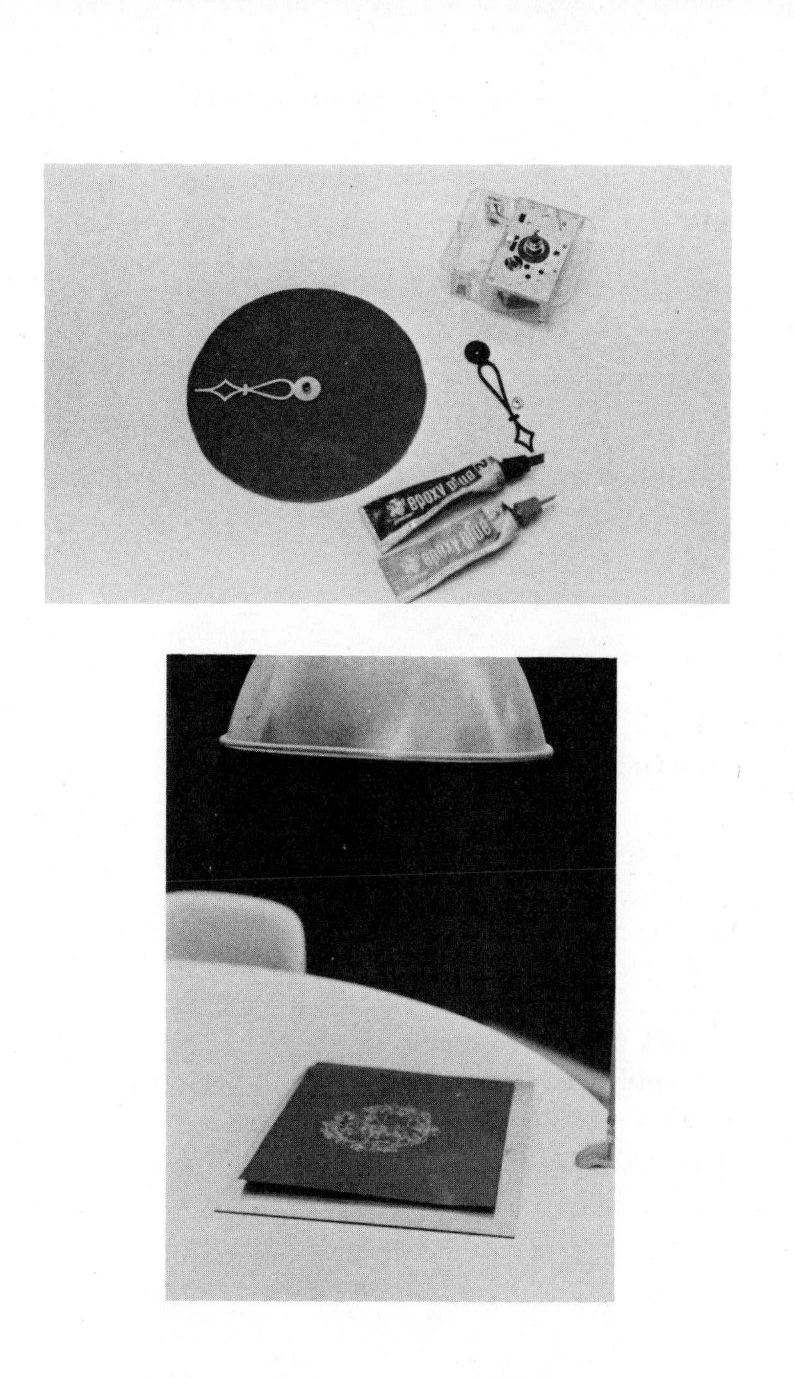
epoxy glue
epoxy glue

MATERIALS

1 deep cigar box for base
2 square feet of decorated, adhesive-backed plastic (Contac or equivalent)
1 flexible gooseneck
6 feet of electrical "zip" cord
1 clock motor with hands
1 sundial disc—decorated plastic or light metal
1 bottle of Pliobond cement
1 pointer—use scrap material

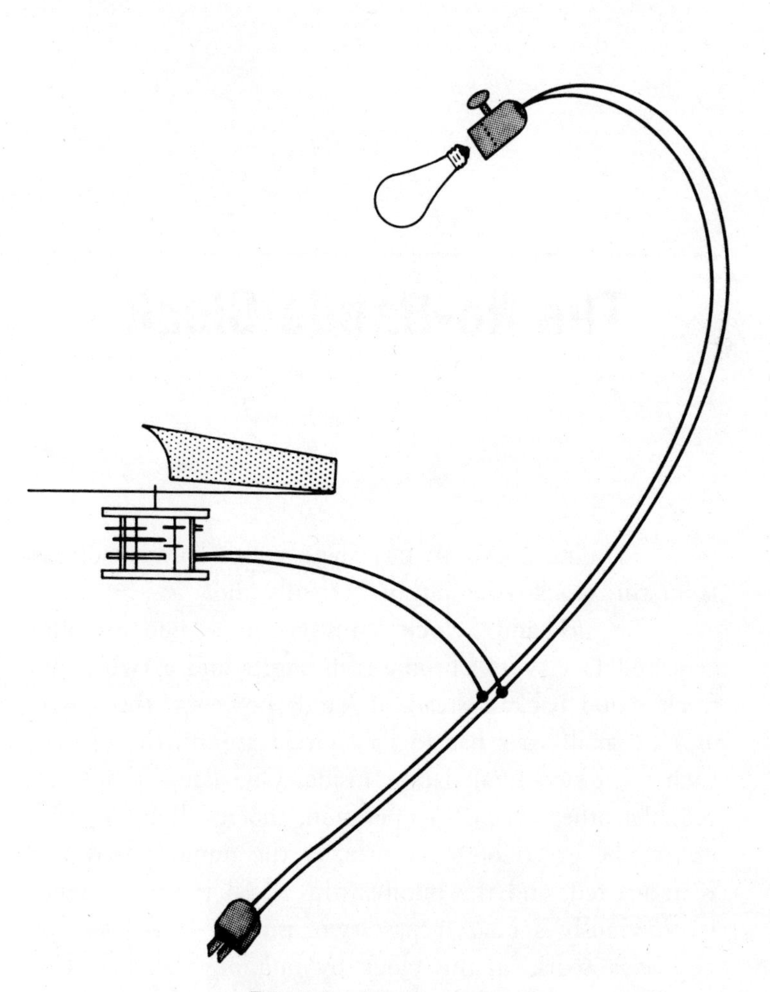

Electric Sundial Wiring

PROJECT 6

The No-Hands Clock

YOU'RE GOING TO GET some very funny reactions from this clock—the author certainly did!

The no-hands clock consists of a fine walnut-veneered face, with chrome-trim angles and a two-inch-thick wood back. Instead of hands however, this clock uses 12 small vials placed in a circle around the center, each with two small lamps inside. One lamp is colored red, the other green. In operation, the red lights signify hours, the green ones minutes. If the number two vial is lit up red, and the number three vial is lit up green, it's obviously a quarter past two, isn't it?

Start work on this clock by building the face. Use a good grade of walnut plywood, one quarter inch thick, and 12 inches square. Make the sides for the case of additional plywood, 2 inches by 12 inches. To hold the face to the sides, top and bottom, obtain a length of brushed aluminum corner trim from your building supply dealer or lumberyard.

Now drill 12 holes, using a one-inch wood drill or if one is available, a circle cutter. Place the holes on the 30° centers, and be careful not to mar the finish of the wood.

Make four small gussets of wood. A piece of wood, one inch by one inch by two inches will provide you with two suitable gussets, when the wood is sawed across the diagonal. Trim an additional quarter inch from each to allow these to serve as supports for the back of the clock, which will be installed later on. Cement the gussets in the four corners of the case.

Now obtain the 12 vials you need. Your local druggist can supply you with one-inch-diameter glass vials. The vials should be about one and one-half inches in length. Check them to see that they have no wording on the bottom.

In order to conceal the fact that bulbs are inside the vials, paint the inside with three coats of pearlescent fingernail polish. This gives a translucent appearance to the vials, which are then cemented into the twelve holes from the back of the clock. For a better overall effect, allow the bottoms to protrude about one-eighth inch from the face of the clock.

Now we have to get tricky.

Obtain a four-inch square of printed circuit material. This copper-clad laminate is used in electronic circuits. Using a soft-lead pencil and a compass, scribe four circles as shown in the diagram. Mark these off into sections, as indicated, and then fill in the sections with acid-resisting ink. When this has dried, place the board in the etchant solution, and periodically remove it, flush it under running water, and hold it to a strong light. When all the excess metal has been eaten away, the light will show through clearly.

A bit of turpentine on a soft cloth will then remove the acid-resist, leaving you with gleaming copper pads right where you want them.

Drill out the center hole, large enough for the armature assembly of an electric clock motor to pass through. Attach the clock to the back of the board. You can then cement the bottom of the clock motor to the inside of the clock face, taking care not to get cement on bearings or on moving parts.

You will need 24 six-volt pilot lamps. These are available at most radio supply houses. Using red and green lacquer, color half red, half green. Make sure that all the glass is covered. When these have dried, carefully solder a 10-inch length of no, 24 stranded wire to the base of each lamp, and another to the side, or body contact of each. This soldering is much easier on brass-based lamps than on the aluminum-based types.

Place one red and one green lamp in each vial, and pass the wires through a small hole in the cap of each. To help identify the wires, use black wires for all the body contacts, a red wire for the red lamps, and a green wire for the green lamps. Connect all the black wires together. The copper area remaining around the perimeter of the etched circuit board will serve as a handy anchoring point all around the clock. Simply solder all the black wires to this area.

The inner segmented ring represents the hours, so connect the red leads in proper sequences to the inner segments. To simplify this wiring, drill small holes in each segment, and pass the wires up from beneath, solder and clip off the excess. Make sure that the solder blobs are all near the outer edges.

Repeat with the green wires for the outer segments.

Now connect a wire from a six-volt filament transformer to the common point where all the black leads were mounted. The other secondary wire goes to the body of the clock itself.

Connect the primary wires (through an on-off switch) to the A.C. line, and connect the clock motor to the same line. For added convenience, a small terminal strip can be used.

Electrical contact between the clock motor and the lamps is made from the hands of the clock.

Using a small file, carefully burnish away any oxides under the clock hands, and solder a small piece of spring to each. As the hands turn, they will drag the spring along the contacts and change the lighting sequence as the time passes.

To set the clock up, plug it in and move the hour hand until the proper red light comes on. Now wait until the time reaches the half hour, and then switch to the next hour spot. Set the minute hand until the half-hour, or number six green light comes on. You see, with this clock, we want to change the hour on the half-hour, as this will make the clock easier to read. Most of us are more prone to say "It's twenty to three," instead of "It's two forty."

This clock operates all the time, but to save wear and tear on the lamps and transformer, the on-off switch permits you to shut the lights down when the clock isn't in use. Turn the switch on again, and the clock gives you the correct time.

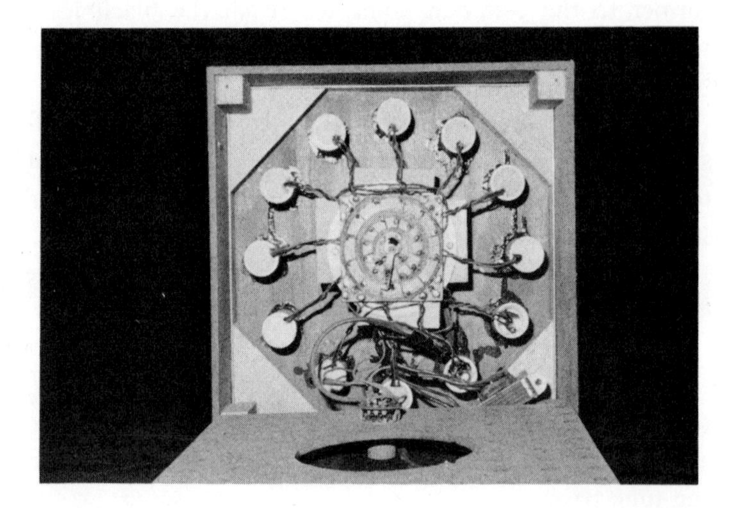

MATERIALS

1 piece of 12-inch square ¼-inch walnut-veneered plywood
4 2-inch by 12-inch ¼-inch walnut sections
1 8-foot length of brushed aluminum corner molding
4 wood gussets (see text)
12 1-inch-diameter glass vials
1 bottle pearlescent fingernail polish
12 6-volt pilot lamps (8-volt will also work)
1 set of glass-coloring lacquers
1 4-inch square of copper laminate, with acid-resisting ink, etchant, and turpentine
1 117 to 6-volt filament step-down transformer
1 electric clock motor with hands
1 on-off toggle switch
1 6-foot cord and plug set

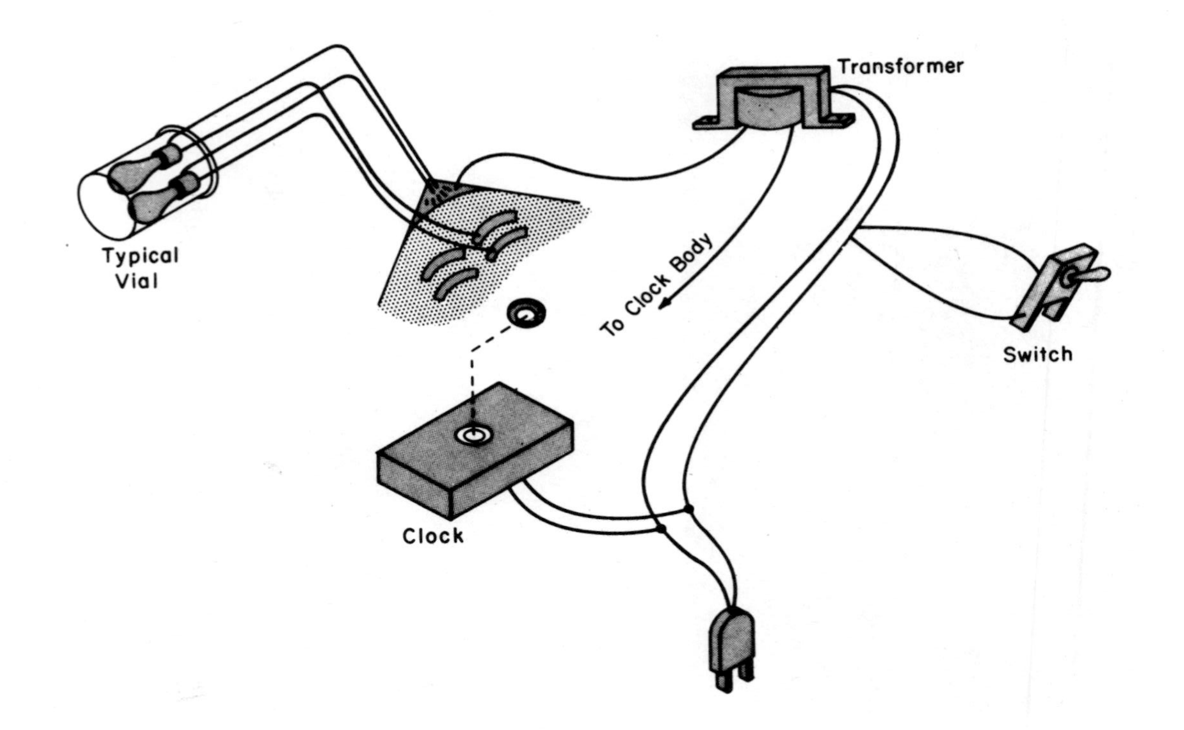
Transformer
Typical Vial
To Clock Body
Switch
Clock

PROJECT 7

The Wall Clock

THIS IS CALLED a wall clock, for it is indeed a clock that is built on, around, and in a wall. It's a good project for those of us who are not particularly handy handymen, as it is easy to construct. With a little imagination, you can design the clock any way that you like.

Pick a nice, clear wall in your home, and make a hole in the wall, adjacent to a vertical stud. The hole should be about four inches in diameter. Once you have the hole, attach a gem box or electrical fitting box to the stud. The top of the box should be on a level with the surface of the plaster.

Place a transistorized clock movement into the box and fasten it there. Make a cover plate using a disc of opaque plastic, one-quarter inch thick. Drill out a center hole to allow clearance for the clock motor armatures.

Attach the plate to the gem box. You will find that the plastic plate neatly covers the hole in the wall.

The hands for this clock are made from wire which is pulled taut to straighten it and is then cut to the proper length. You can paint the wire hands to suit your own tastes. Attach the wire hands to the clock's friction-fit hands with either solder or a good grade of epoxy cement, and fasten the hands to the armatures.

Here's where you can get creative.

For time indications, use wood, plastic, or metal markers—whatever you want—and cement these markers directly on the wall. At the end of this book, you will find sources for such materials, and you will have no difficulty in locating something suitable to use as markers.

MATERIALS

1 electrical junction box with oversize cover plate
1 transistorized clock movement
Assorted wood, plastic, or metal markers used for time indications
Oversize hands made from wire (see text)

PROJECT 8

The "Spider and the Fly" Clock

THE "SPIDER AND THE FLY" clock is a really unusual beauty and very easy to assemble.

The effect is that of a plastic sheet framed in a wooden box. The front of the plastic sheet is decorated with a wispy spider web, around which a small fly tries to escape from a small spider. You tell the time by observing the positions of the two.

To construct this clock, start by making the box housing. You can use a cigar box and cover it with adhesive-backed decorated plastic. Remove the box top first, and install triangular wooden gussets, or short lengths of wood dowel at each corner. These will support the plastic face of the clock.

Mount the clock so that the hands will be just *below* the plastic face. Do not remove the clock hands.

The plastic you select should be opaque. By referring to the sources in the back of the book, you will find assortments in dark blue, black, red, or whatever color you fancy. The plastic will come with a peel-away paper covering which protects the surfaces. Leave the paper on the back, and remove it from the front. Cut the plastic sheet so that it fits into the box and rests on the corner gussets.

With a sharp scriber trace a spider web on the plastic face, scratching the lines into the plastic, When this is finished, rub paint filler into the lines, and wipe away the excess. Allow this to dry thoroughly before proceeding. If you have selected a dark-colored plastic, use white filler; if you choose a light-colored plastic, select a dark filler.

Now place a small alnico-V magnet on each of the clock hands, and if necessary, bend the hands slightly to be sure the minute and hour hands will clear.

Cut a small fly from a magazine illustration, and find a spider the same way. Cement a small straight pin to the underside of each. Now place the spider over the hour-hand magnet, the fly over the minute-hand magnet. The magnets will engage the pins, and once the clock has been set and plugged in, the fly and spider will rotate in concentric circles over the face of the clock.

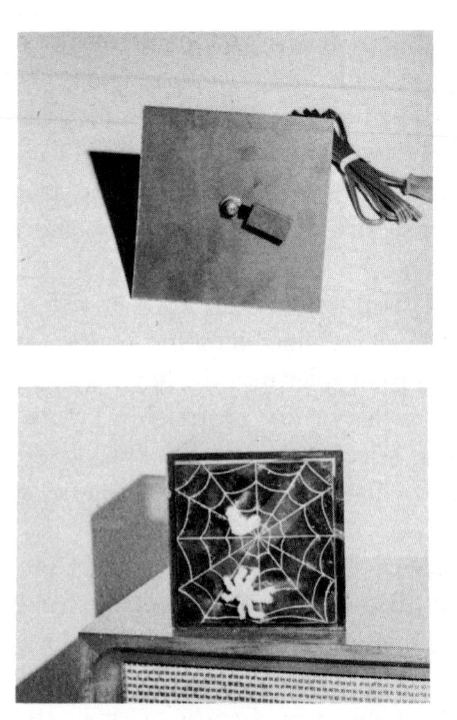

MATERIALS

1 large cigar box, covered with decorated, adhesive-backed plastic
1 complete clock mechanism with hands
4 short lengths of dowel for gussets
1 sheet of opaque plastic for clock face
1 stick of filler enamel
2 small alnico-V magnets
2 small straight pins
1 fly and spider, cut from magazines

PROJECT 9

The Pendulum Clock

A PENDULUM CLOCK, be it housed in a grandfather's case or a simple wall-mounted cuckoo, is a thing of beauty. In the early days, the clock was wound by pulling heavy brass weights to the top of the clock case. As their own weight drew them down, energy was imparted to the escapement wheel, whose rocker arm, or pawl, was controlled by the swinging of the pendulum. Raising the pendulum would shorten the moment arm, speeding the clock, and lowering this weight would have the opposite effect.

Today, however, we have a vast choice of highly accurate clock movements, and they operate from electric power. We don't really need pendulums anymore, but people have come to miss the warmth and comfort supplied by the old tick-tockers and their slowly swaying pendulums.

This project reveals a simple ruse to emulate the grace and charm of yore using a modern electric clock. The pendulum has nothing whatever to do with keeping time—in fact, it is totally separate from the clock itself! Once you understand the principle applied here, you can package your own pendulum clock any way you choose.

To start, you need a standard electric clock fit-up, consisting of a clockwork mechanism, glass bezel, clock face, and hands. This is the time-keeping device.

Below the clock motor, mount a second motor, which has an eccentric cam drive fixed to its shaft. You can make the cam by simply bending a piece of ordinary clothes hanger wire around a screwdriver blade, and then clipping it with a heavy-duty wire cutter. A slip-joint pliers will help you bend the cam to the desired shape.

Another length of hanger wire serves as the pendulum arm. You can make this as long or as short as you wish. If you want to build a grandfather's clock that stands on the floor, you'll need more length than a clothes hanger can provide; any piece of metal rod will do.

You will need a pivot point below the clock to support the pendulum rod, or you can (if you choose a clock with a long bushing) suspend the rod from the clock bushing. In any case, bend the upper part of the rod over and do not clip it (see diagram). Now shape the rod as shown, so that the cam will be free to move up and down between the two wire parts.

As you can see, starting the lower motor causes the eccentric cam to revolve, and as it turns, it moves freely up and down between the wires. The rotation of the eccentric cam causes the pendulum rod to move rhythmically from side to side.

A pendulum weight is also necessary for the right appearance. In this case, the weight just "goes along for the ride," so the less weight, the better. Take the cap from a large jar and drill it at one side so the rod can be slipped into place. A drop of cement will hold it solidly.

Give the pendulum and the exposed part of its rod a coat of gold paint to finish the job.

In mounting the system, be sure that the covering face of the housing extends below the pair of wires, so as not to reveal the mechanism. You can make many attractive housings for a clock such as this, and the choice is limited only by your imagination and your ability as a cabinet-maker.

For realism, add a little tick-tock, You can easily build a tick-tock into your pendulum clock by simply adding two small wood blocks to either side of the pendulum swing. Make the blocks long enough to reach from the sides of your housing to the point at which they will be contacted by the pendulum rod at each limit of swing.

MATERIALS

1 complete clock fit-up, Lanshire or equivalent
1 timing motor (Edmund or equivalent)
1 coat-hanger wire, cut and bent per diagram
1 can or jar lid for pendulum weight
1 housing (decorated cigar box)

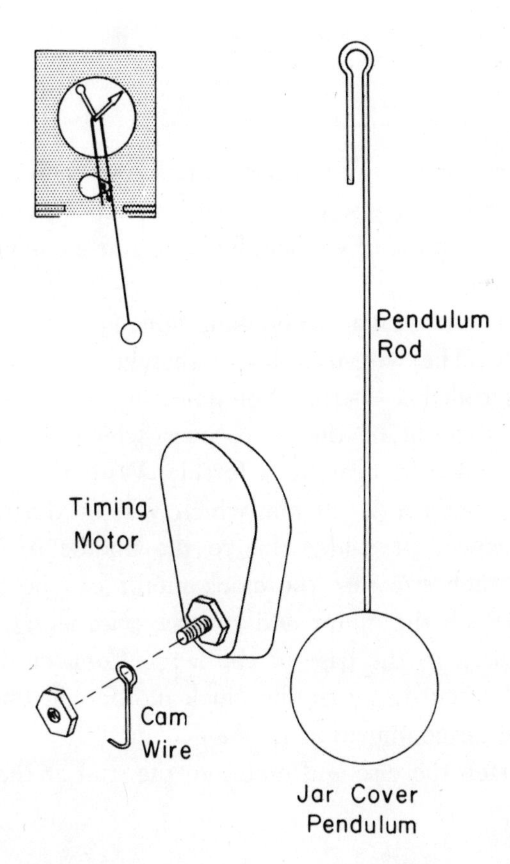

PROJECT 10

The Lamp Clock

SPACE IS OFTEN at a premium, especially in urban apartments. You just cannot devote such valuable table space to a lamp *and* a clock, but this unit combines both, very simply.

Start out with a nice-looking table lamp, and remove its shade. The wire-like device that holds the shade in place is called a "harp." You'll usually find a threaded stud at the top of the harp, over which the shade is placed, and held firm by a ferrule. With the shade removed, obtain a ¼-20 nut which will fit the threaded stud. Cement or solder this to the middle of a metal disc of such size that the clock motor can be attached to it. Attach the motor and run the wire along one leg of the harp to the base of the lamp. Connect this wire to the line cord so that the clock motor will operate as long as the combination is plugged in.

Attach the disc and motor to the stud at the top of the harp.

Now obtain a ¼-20 machine screw with a flat head. Thread this into two ¼-20 hexagonal nuts, tightening one against the other. Now you can lock the nuts in a vise, and the screw will not turn. Drill a hole into the top of the screw, about one-quarter inch deep, and one-eighth inch in diameter. Carefully solder or cement the clock's hour hand to the head of the screw, making sure that the holes are centered.

When the hand is placed in position on the clock, the concentric minute and second armatures will slip into the hole, and rotate freely without binding.

Now go back to the lampshade, and using black India ink, mark the lower rim of the shade in equal increments so that a total of twelve segments are marked off. You can use lighter lines to indicate subdivisions of halves or quarters. Add small numbers to identify the twelve hour markings.

To provide additional accuracy, make a pointer out of thin wire, and affix this to the lamp support in such a way that it indicates the position of the lampshade relative to itself.

As you will see, when you plug this in, the lampshade (which is held in place by the stud on the hour-hand armature) will rotate one complete revolution every 12 hours.

7
6

MATERIALS

1 table lamp with harp support
1 clock mechanism with hour hand
3 ¼-20 hex nuts
1 ¼-20 x ¾-inch flathead machine screw
1 bottle of black India ink
1 wire pointer (see text)

PROJECT 11

The Personalized Fun Clock

REMEMBER THE OLD Mickey Mouse wristwatches? Well, we're going to make a Mickey Mouse clock—with you instead of Mickey Mouse on the face!

Start by taking a good-quality photograph of yourself (or have a photographer friend take one for you) wearing an all-white outfit on a black background. Wear a long-sleeved white shirt, preferably a tight-fitting one which will be easier to cut out in the photograph later, and white slacks and shoes—sneakers will do nicely.

Have two prints enlarged. One should be sufficiently large to fill the face of a good-sized clock. Make sure that the prints are not ferrotyped, but are dried flat.

Using black opaque ink, carefully paint out the arms on one picture. You will now have an all-black background with you dressed in white in the middle of it. Cut this photo out so it fits the bezel of your clock.

On the other photo, carefully cut the arms out of the picture, and cutting from the shoulder, make one arm slightly shorter than the other, This will be for the hour hand.

Carefully remove the hands from the clock, and mark the exact center of the photo. Cut a small hole to allow the clockwork armatures to pass through. Now cement the photo to the clock face.

You can paint the numbers on the black background with white paint, or, should you prefer a classier job, obtain white rub-off type and put the numbers on this way.

Cement the arms to the top surfaces of the clock hands. You will have to provide a hole in the longer minute hand to pass the armature of the hour hand, which is then put in place.

And there you are—Mickey Mouse!

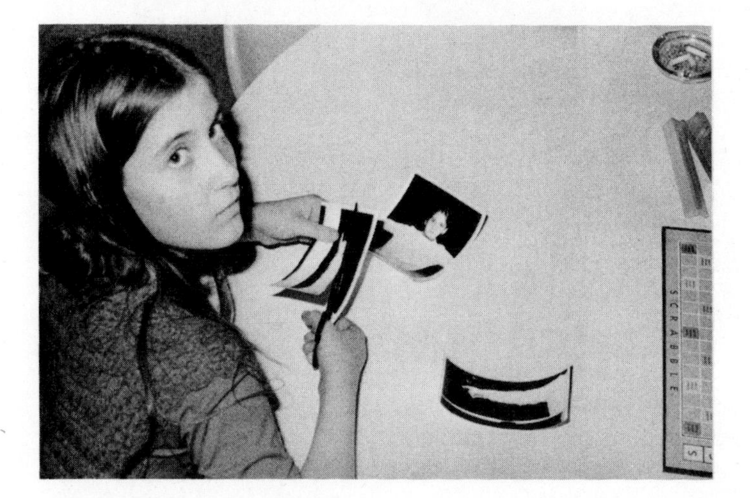

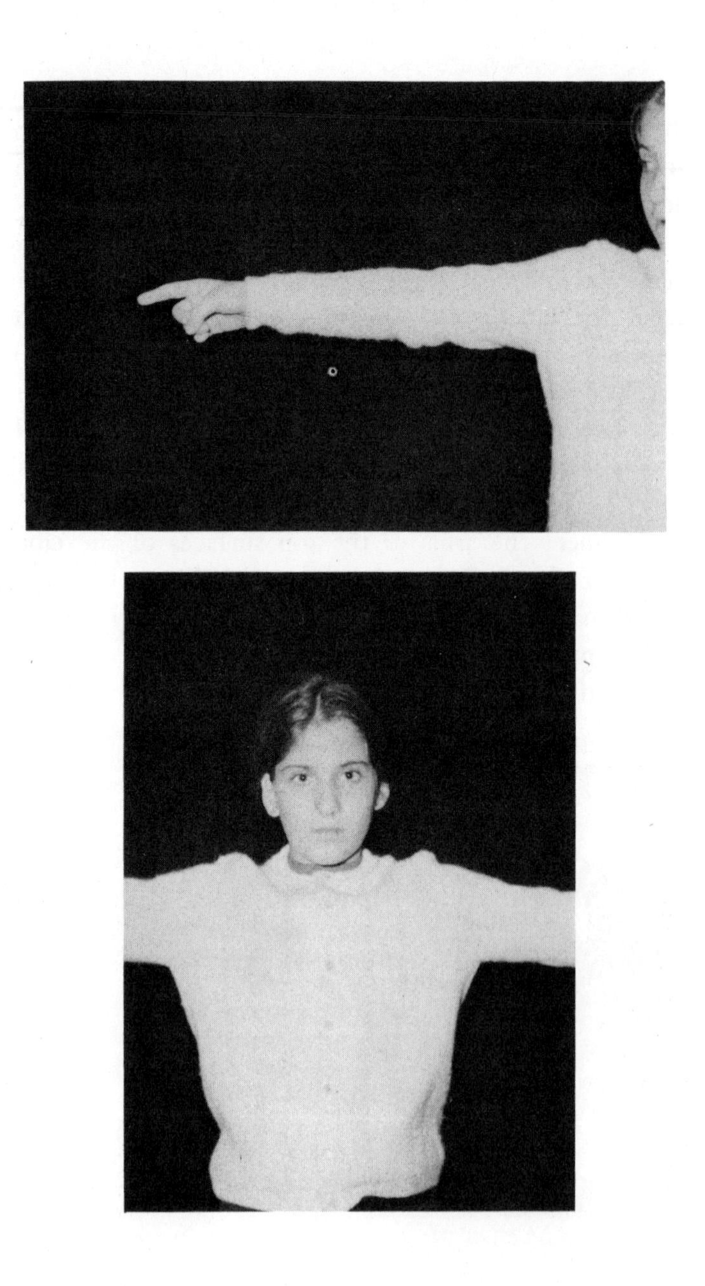

MATERIALS

1 electric clockwork with frame, bezel, glass, and hands
2 photographs of yourself (see text)
1 bottle of opaque ink
1 set of rub-off type in white

PROJECT 12

The Cube Clock

CUBES ARE "IN." Plastic cubes, used as points of interest through the house, are now high-fashion accessories. And this high-fashion clock will tell you the time—in a most interesting manner.

The cube itself is constructed of a dark-colored translucent plastic. You can obtain some very pretty plastics from which to construct your cube—deep blues, smoke colors, dark reds.

But first, let's look at the effect of this clock.

It's a beautiful square cube, and if you look at the front face, you'll see three small points of light. One, the movement of which is too slow to really be seen, tells you the hour; the next, which moves faster, tells you the minute; and finally, the third, which moves around the face of the cube at a rapid rate, tells you the second.

Obviously, the problem in this clock is one of commutation, for you need a way to bring the current to the lamps and allow the hands to rotate freely. We've solved the problem here.

Start by selecting your plastic and get it in at least a one-eighth-inch width. One-quarter inch will give you an even more rigid box to work with. Leave the back of the cube open.

Make the cube about eight inches square, and cement the sides, bottom, top, and front with a high-quality acetone-base plastic cement. Do not overdo the use of cement, as it will run and mar the surfaces it touches. When the cube has dried, you are ready to proceed.

Obtain a six-inch square piece of copper-laminated phenolic, such as is used in etched circuit work. Drill a three-eighth-inch hole in the center, to pass the clock armature. By etching, remove the copper from the board near this hole as well as near any of the mounting holes with which you will then attach the clock to the back of the board. The clock body should not touch the copper at any point.

Carefully scrape any paint or oxides from the clock hands, and bend the pointer ends of the hands at 90 degrees. Tin these with solder, and solder the body contacts of three 6-volt lamps in place, one at the end of each hand.

Now obtain the spring-like contact arms from an old electric relay, and clip these to their full length. Bend the contact points at right angles, and solder these to the base contacts of each lamp, making them just long enough to touch the copper surface when the hands are installed on the clock armatures.

Now solder one end of a ten-inch long wire to the edge of the copper surface of the mounting board. It will be connected later.

Solder two small angle brackets to the bottom of the board, and mark the plastic bottom of your cube to locate the holes. Drill the holes out carefully and countersink from the bottom. Use two flathead machine screws, lockwashers, and nuts to mount the board (and the clock) vertically on the bottom.

Using additional flathead screws, mount the 6-volt transformer to the bottom of the cube. Connect the secondary wires so that one joins the ten-inch wire from the copper strip while the other connects to the body of the clock. The primary wires connect to (1) the power cord and (2) through an on-off switch to the power cord.

The wires to the clockwork mechanism should be connected directly across the power line, before it goes to the switch,

Now remove the bracket and take the clockwork system out of the cube. Set it about one-half hour ahead of the actual time, and then reinstall it. When the time has properly caught up, plug the clock in and turn the switch on. Look at the face of the clock, and you should be able to see the lights and tell the time by them. To save wear and tear on the bulbs, turn the switch off when the clock is not in actual use. The lights will go out, but the clock will continue to tell time accurately and be right on time again when you turn the switch back on.

If you feel the need for additional accuracy, don't hesitate to scribe a circle (a five-inch circle should do nicely) on the face of the cube, and then add radial lines scribed with the aid of a ruler to indicate the five-minute segments of the face of the clock. A bit of white stick-enamel will make these lines stand out.

MATERIALS

5 pieces of ¼ x 8 x 8-inch translucent plastic
1 tube of plastic cement
1 electric clock mechanism
3 6-volt radio pilot lamps
1 old electrical relay
1 piece of 6-inch-square copper-laminated phenolic
1 SPST electrical switch
2 small right-angle brackets
4 flathead machine screws with nuts and lockwashers
1 6-volt filament transformer

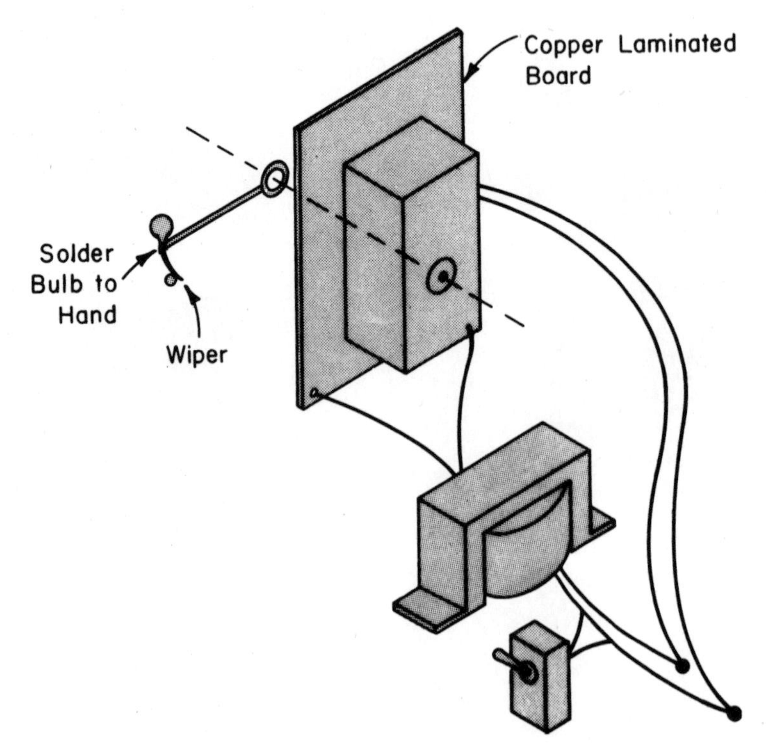

PROJECT 13

The Sphere Clock

IN THE MODERN IDIOM, high fashion in decorating means the tasteful and artistic use of shapes. The "in" shapes these days are cubes and spheres, used about the room as points of interest. However, these shapes *must* be functional. So cubes become record storage cabinets, and spheres and cubes become lamps, and before you finish stacking and rearranging, your room takes on a mod look.

Here's a functional sphere that becomes a proud addition to your home, for it's a combination that serves as a light, a clock, and a definite point of interest as well.

To make this unit, you're going to need some fancy plastic work, and you'd be best advised to have this done professionally. Try a local sign maker for the material and/or the work on the housing. Remember, the more difficult it is to come by, chances are the more unusual and original the unit in your home will be.

You'll need two plastic hemispheres, one in clear plastic, another in frosted white. They should be equal in size, so that when placed mouth-to-mouth, they become a complete sphere. Six inches in diameter is a good all-round size to start with,

Have the two spheres cut about two inches up from the mouth, so the tops are totally removed. Then have the cut edges polished again.

Using plastic cement, start by cementing the two larger pieces together, mouth-to-mouth as they were before. We now should have a large clear ring cemented to a large frosted ring. Now carefully cement the smaller frosted dome to the open end of the clear ring.

Mount your electric clock mechanism to a clock face made of thin, rigid plastic. The plastic should be opaque; considering the materials we've been working with, white would be an excellent color for this particular clock. Mount a small lamp socket to the clockwork frame, orienting it so that it will be near the clock, but not interfere with the operation.

Place the plastic clock face with the mechanism inside the sphere, and adjust the lamp socket so that the face fits flush on the sphere. To make sure that it will fit, also install a small lamp (not over 15 watts) in the socket.

Now mark the underside of the clock face with a soft pencil, using the edge of the sphere as a guide. Remove the face and trim the excess with sandpaper. Check the fit periodically to see that you do not remove too much.

Drill a one-half-inch hole in the lower frosted section of the sphere, and install a plastic grommet for the line cord. Another half-inch hole just above this will do for the on-off switch for the lamp.

Install the switch. Pass the line cord through the grommet, and connect one wire to the switch, the other to one contact of the lamp socket. Connect a short length of wire between the open switch terminal and the other lamp socket contact. Connect the clockwork mechanism wires across the line cord. The wiring is now complete,

Very carefully, apply a small drop of plastic cement to three equidistant points on the open end of the sphere, and put the clock face into position on the open sphere. Allow the cement to dry thoroughly.

Now apply a thin layer of cement to the clear plastic dome that remains, and cement it to the top of the clock face. When it is dry, you will have a sphere, consisting of a clear plastic top which reveals the clock face, a frosted section, followed by a clear section, and a frosted dome at the bottom.

Make a small three-legged stand out of brass wire, forming it like a chemical ring-stand. Place the sphere in the stand so that the frosted dome is at the bottom. Plug the clock in, and your time-telling lamp-sphere is ready for use.

It should be explained that because of the design of this unit, you will have to wait until the correct time is actually reached to plug the unit in and have it read with accuracy. If you must reset this clock, or change the lamp, use a sharp razor blade and cut the clock face from the frosted ring. Change lamps, or reset the clock, and then recement it as before.

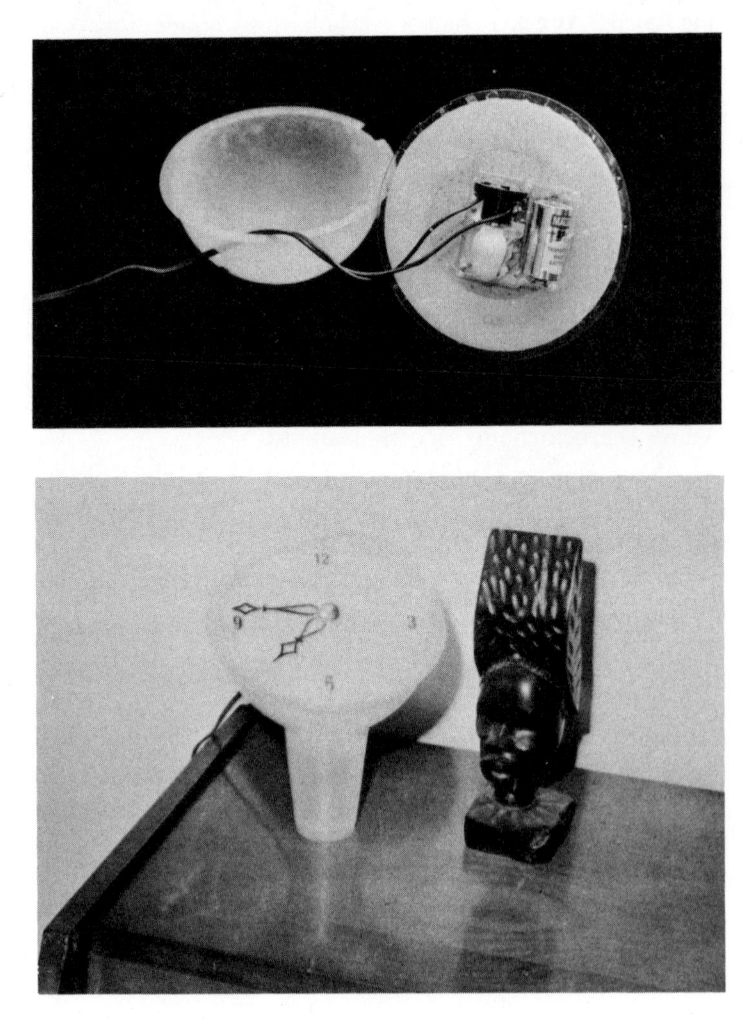

MATERIALS

2 plastic hemispheres, one frosted, one clear, 6 inches in diameter
1 electric clockwork mechanism
1 white plastic sheet, about 5 inches in diameter
1 tube of plastic cement
1 small lamp socket
1 15-watt lamp
1 plastic line-cord grommet
1 6-foot plug and cord set
1 SPST switch
1 length of brass wire

PROJECT 14

The Decorator's Special

THE DECORATOR'S SPECIAL CLOCK does a lot more than just tell you the time. The face constantly changes patterns, making it a fine decorator item. All it takes is a little ingenuity and a steady hand with a paint brush.

Mount your clockwork mechanism to a white matte plastic disc, which will serve as the face of the clock. Make this face about six inches in diameter. Drill a hole through the center to pass the clockwork armatures.

With a soft pencil, mark a circle one inch in from the outer perimeter of this disc, and put the clock numbers around this circle. Use Roman numerals, in a rub-off type.

Now draw thin lines, extending in a slow spiral from the center of the disc. These lines can take on any number of patterns, from lightning-like jagged patterns to staircase designs, or simply easy-flowing lines. Use India ink to darken them when you achieve a pattern that you like.

Now make another disc of lightweight but rigid clear plastic. One-sixteenth-inch or one-eighth-inch plexiglas would be ideal. Scribe a one-inch circle around the outer perimeter of this six-inch disc, and repeat the design in the opposite direction, using the India ink again. Incidentally, you should have no difficulty in getting the ink to adhere and "wet" the matte surface, but on the clear plastic it will tend to ball up. You can easily overcome this by adding a few drops of liquid detergent to the ink. This acts as a wetting agent and reduces the ink's surface tension.

Cement the hub of the clock's second hand to the middle of this disc, and after putting the hour and minute hands in place, put the disc on.

Now mount your clock in a suitable box, and plug it in. As the hands give you the hour and minute, the rotating disc causes an interplay of patterns on the face of the clock, one that seems to be constantly in change, and never the same twice.

9
3

MATERIALS

1 6-inch white matte plastic disc
1 6-inch clear plastic disc
1 clockwork mechanism
1 cord and plug set
India ink

PROJECT 15

The Cocktail Clock

"DRINK BEFORE FIVE? Not me!"

For people who won't touch a drink until the cocktail hour, this clock solves the problem. No matter where the hands point, it's always five o'clock!

The method used in preparing this clock can be easily adapted to any of a number of varieties of clock. Once you have selected the clock you are going to work with, visit your local art supply store and look at the assortment of rub-off type. You'll find numbers in all sizes, shapes, colors, and languages. You can make a clock with Chinese numerals, if you like.

To start with, remove the clock face by carefully lifting the tabs of the bezel and then slip the bezel and glass from the face of the clock. Remove the hands by grasping at the hub and pulling.

Now remove the clock face and turn it over. Use

a can of white spray paint and paint the *back* of the clock face, so you can easily reverse it, You can simply spray the front if you wish, but make sure to use a totally opaque spray.

When the paint has thoroughly dried, use the rub-off type, and put a number "5" at each point around the clock where a number should go. Now reassemble the clock, and there you are—a cocktail-hour clock.

MATERIALS

1 clock, case, and movement
1 sheet of rub-off type
1 can of opaque white spray paint

PROJECT 16

The Kit Clock

IN THE LAFAYETTE CATALOG, you'll find a novel clock in kit form. If you don't have a catalog handy, the ordering number is 19E31054 and the price is only $11.95. In addition to providing the local time, the clock has a rotating outer face that enables you to read the time in any city in the world. This clock is interesting, educational, and guaranteed to absorb any youngster. But be prepared to have some answers ready when you're asked about "Greenwich Mean Time."

You'll find that the simple instructions enable you to put this almost all-plastic kit together with little or no difficulty, and setting the clock up to operate is simple. The clock comes complete with plastic decorative pieces, including the numerals and face markers, which are gold flashed.

In assembling mine, I was chagrined to find that there

was no black plastic paint available. Making do with what I had, I painted the inside of the clock body with a tempera, which proceeded to crack and cockle when it dried. Damning my own stupidity at first, I started thinking about how to undo the damage. Upon examining the face however, I realized that this unusual effect enhanced the appearance of the clock, so I sprayed it with a fixative to hold the paint properly in position—another case of a successful accident!

PROJECT 17

The Propeller Clock

THE PROPELLER CLOCK is one of the author's favorites and will no doubt appeal to anyone who loves airplanes and flying. Start out at a very tender age with a burning desire to fly. Spend your early years building model airplanes and flying them. Also, go to every movie you can find that has anything to do with airplanes.

Spend your weekends and the time after school hanging around your local airport, sweeping up, helping line boys, and getting half drunk listening to the "hangar bums" as they talk flying talk.

Then enlist in the "Army Air Corps" in World War II, and get yourself sent to China-Burma-India, where you find a Japanese Zero Fighter plane that had been shot down. It's already been hacked at for souvenirs, but the propeller is intact. Imagine how that will look back home on your bedroom wall!

When you get time off, you get busy under the broiling Burma sun in shorts and a pith helmet, working away with wire cutters, wrenches, and band-aids. Finally you get the prop dismounted, only to find that it's far too heavy to move, and if you tried to hang it up, it would wreck the bedroom wall.

Then the war ends, and all about you, material is being destroyed so it won't be taken home. And you find a perfectly good wooden propeller from an L-5 Observation plane . . . And you can almost lift it with one hand.

When the engineering officer points out that you can't have it 'cause it's in good condition, go at it with a hammer and chisel to put a split in the hard, laminated wood.

Then get it autographed by all the officers and men in your squadron so it becomes a legitimate souvenir, and then carry—yes *carry* it all the way home!

With a burning pencil, burn the names in to stay. Then twenty-five years later, write a book about clocks, and you're ready to start.

A visit to your local airport may turn up an old propeller. If it's shattered, that really doesn't matter.

Use a saber saw to cut the hub area to fit your clockwork mechanism, and set the clock in place. You'll find it more expedient, incidentally, to cement this clock into position than to use the mounting nuts. Silastic RTV sealant works well for this, as it remains soft and rubbery yet holds the mechanism firmly. Just be careful not to get the sealant into the clockwork.

To hang the propeller clock, make a hanger of a

length of picture wire, and run it into the backs of the propeller-boss mounting holes and wedge and seal the ends with the Silastic sealant.

The propeller clock makes a handsome and unusual addition to the decor of a den or office.

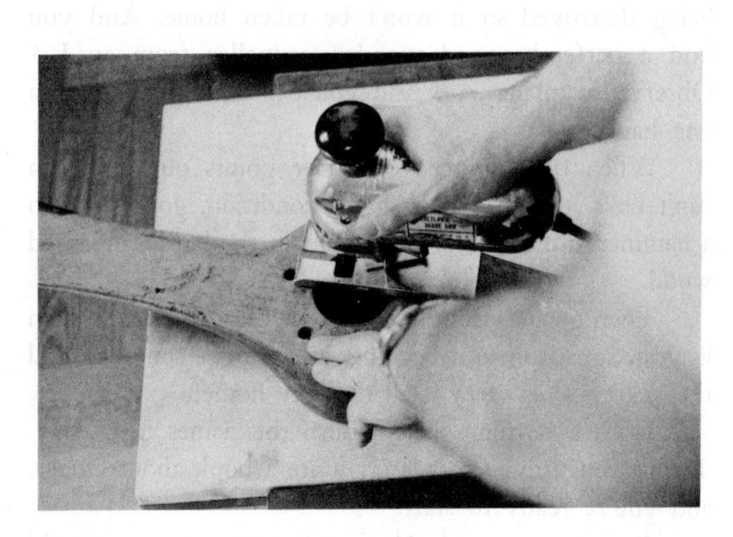

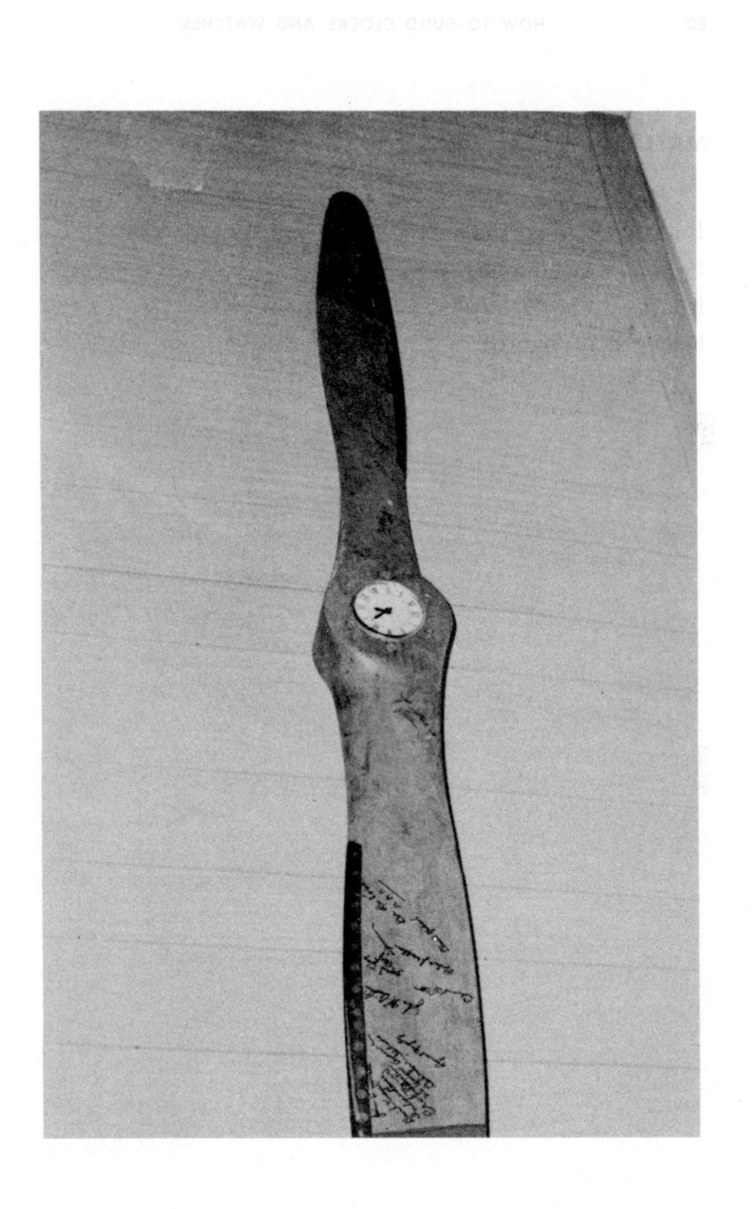

MATERIALS

1 airplane propeller
1 clockwork mechanism
1 large tube of Silastic RTV sealant
1 length of picture wire

PROJECT 18

The Gourmet's Kitchen Clock

PEOPLE LOOK UPON SNAILS, or escargot, in one of two ways. Either it's "Umm, garlicy delicious," or it's "Yuch—people really *eat* those slimy things?"

For those who think snails are delicious, making this clock is a real treat. Not only is the clock an unusual and attractive conversation timepiece for the kitchen or dining room, but when you buy the snail shells that go on its face, you can get them with a small tin of imported french snails. Prepare these with a thick sauce of garlic and butter, and you can enjoy the snails while everybody else enjoys the clock.

A local Japanese store yielded a large-diameter plastic serving dish, yellow on the front and black on the back. The plastic platter was selected as it is easier to drill than a crockery type. Locate the center of the dish by measuring the diameter and marking the radius at several points.

Drill a one-eighth-inch pilot hole in the center, then a one-quarter-inch hole which can be reamed out to a diameter of three-eighths of an inch to pass the operating core of a clockwork mechanism.

Put the clock into place to check the hole size, and then remove the clock temporarily. The shells are more easily assembled without the clock on the back of the platter.

The shells can be cemented into place or attached with double-sided tape. Cut a small piece of the tape, apply it to the back of a shell, and then press the shell into position on the platter.

With the shells now attached firmly to the platter, put the clock in place, and tighten the mounting nut. If you find that the plastic around the hole is rough, you can chamfer the hole edges top and bottom by running a one-inch diameter drill around the hole on both sides.

Escargot are normally eaten with a small fork, a small pick, and a spring-like device to hold the hot shells. Use a small fork for the minute hand of the clock, and a pick for the hour hand.

To attach the hands, drill a hole in the handle of the pick so that the shaft of the minute-hand drive can pass through. Then cement the clock's hour hand to the back of the pick so it can be attached to the shaft of the hour-hand drive. Next, cement the clock's minute hand to the back of the fork so that it can be put in place. The second hand is not used on this clock.

Assemble the pick and fork to their respective shafts, and put the battery into the transistorized clock movement. Set the time, and the clock will keep time for you.

To hang this clock up, attach a small hook on the back of the clock with a dab of strong epoxy cement.

MATERIALS

1 transistorized clock movement and battery
1 lightweight, large-diameter plastic serving dish
12 cleaned and dried snail shells
1 roll of double-sided tape
1 set of snail implements: fork and pick

PROJECT 19

The Hi-Fi/Stereo Time Control

SOME CLOCKS CAN BE MADE to do more than simply tell the time. This clock goes to work for you in a way that you can use, and use right now.

Suppose you have a complete stereo system, and are fond of tape recording directly from on-the-air FM programs. And while we're supposing, let's assume that you have an urgent appointment one evening, and while you're keeping your date, a much-wanted recording is going to be aired.

Are you going to miss it? Will you pass up the chance to tape this recording, just because you aren't there? Not with *this* set-up, you won't!

The clock used here is the type that is used in many clock radios. The feature we're most interested in is the switch that allows you to set a time or alarm for waking you up in the morning. Instead of using the switch to turn on the electric alarm however, we use it to control a relay which in turn will activate the FM tuner, the amplifier, and the tape recorder, all of which have been preset to the proper levels and positions.

Now when you are ready to leave your home, set the timer and the controls, allowing about three to five minutes in advance so the system can warm up. Turn everything on to make sure it's all set, and then activate the clock. You leave home, and at the proper time the recording is made. You come home, rewind the tape, and there it is, all ready for editing or just listening to.

To make this set-up, cut a hole in a metal panel to suit the clock motor, and mount the clock. Obtain a two-section, double-pole double-throw relay with a 117-volt coil. Mount this on the back of the panel. For added convenience, wire a group of three outlets to the power line through the relay contacts.

Now simply plug the amplifier, FM tuner, and tape recorder into this relay-controlled outlet, and that's all there is to it. Plug the clock meter into a live outlet, and you're all set.

If your recorder is equipped with an end-of-tape cut-off, tape into this line for a second relay to turn things off by breaking the first relay's power line (see diagram).

If your recorder does not have an end-of-tape cut-off, this may mean that after the tape runs out, your recorder will continue to operate until you come home and turn it off.

To avoid this, you can easily make an end-of-tape cut-off. Obtain a small microswitch with a leaf-type actuator and mount it in the path of the tape. The tape's weight will keep this switch closed. When the tape runs out, the switch (and the power line to the machine's motor) will open.

Place the actuator leaf of the microswitch in such a way that it engages the shiny, or uncoated, side of the tape.

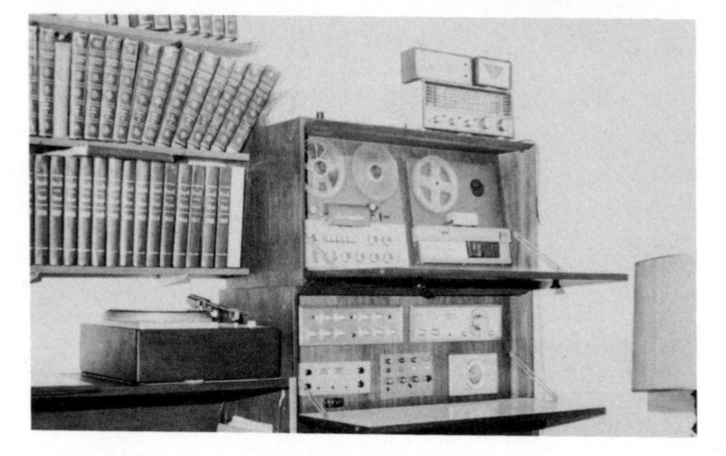

MATERIALS

1 clock mechanism from clock radio
1 mounting panel to suit
1 2-pole, double-throw relay with 117-volt coil
3 A.C. outlets
1 6-foot line cord and plug set

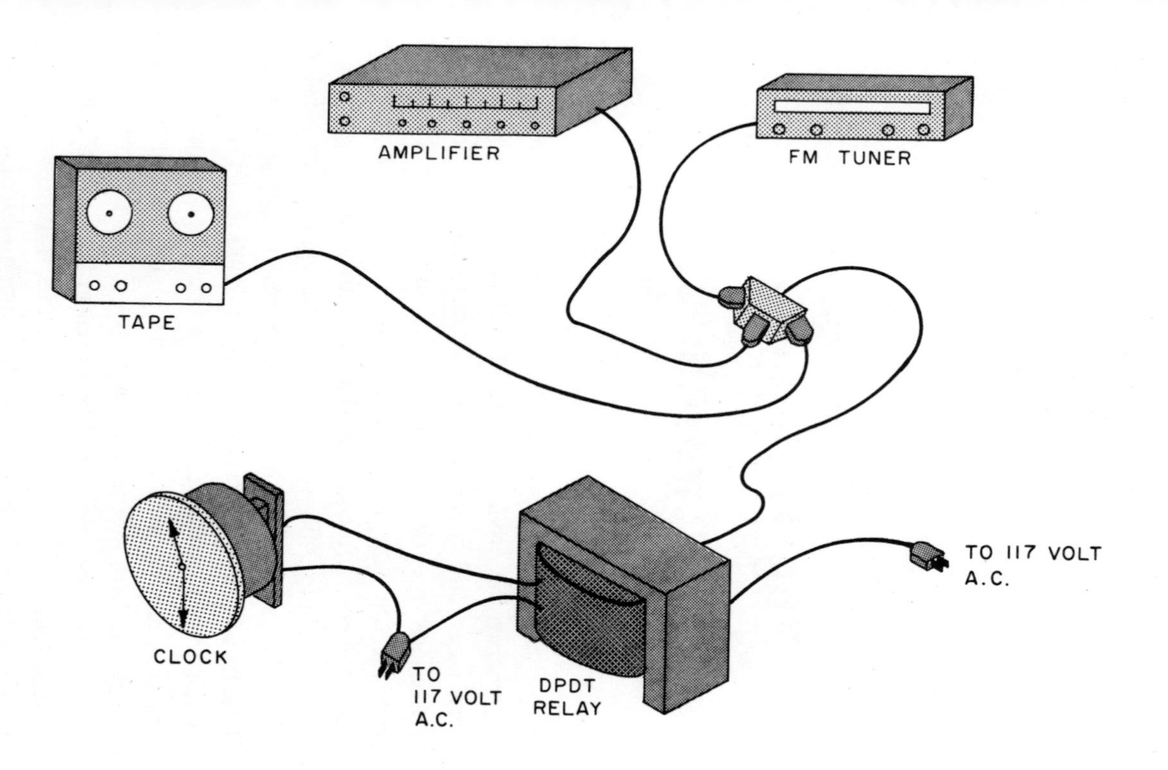
AMPLIFIER
FM TUNER
TAPE
CLOCK
TO 117 VOLT A.C.
DPDT RELAY
TO 117 VOLT A.C.

PROJECT 20

The Traveler's Wristwatch

ANYBODY WHO DOES any amount of traveling will appreciate the obvious values of this handy watch. It has *two* faces, and is actually two watches in one. One face represents the time at home, the other face tells the traveler the time in the time zone he's visiting. If this seems a conceit, consider that it enables the traveler to avoid making calls too late at night, and waking folks back home. What's more, the constant indicator helps him to adjust to time-zone changes, making his travel so much more comfortable. When he's at home, both watches are set at the same time, giving him a double-check for better time accuracy.

There are any number of approaches that can be taken, including mounting two ladies watch movements in a plastic block. However, after much experimentation, the following idea seems the best and most acceptable.

Obtain two watches. In selecting these, it will be an advantage to provide additional accessories such as a calender and stop-watch on one, an alarm on the other. The important thing here is to make sure that the small expansion pins that hold the strap are of the same size,

Half the strap must be removed from each watch. Do this carefully, making sure that the half with the buckle is removed from one watch, the half without the buckle removed from the other.

Now obtain a small piece of aluminum (an aluminum beer can will provide this for you). Cut it with a heavy shears so that it is about three-quarters of an inch long, and just wide enough to slip into place where the straps were.

Use a file to dress off any rough edges that may remain. Now slip the aluminum strap under the expansion pin on one watch, and bend the strap around. Repeat with the other side for the second watch, and carefully peen the strap flat with a small hammer.

Now carefully drill a hole through all three thicknesses of this metal, using a drill large enough to accommodate a small rivet. You can use a "pop" RiveTool to put the rivet into place, and the job is done.

What you now have is a pair of watches, joined by an aluminum band. Place it around your wrist, and you have an unusual and extremely practical wrist watch that will surely pop a few eyes when you pull your sleeve back to "double-check" the time!

The traveler's watch can also be made using only the strap. Remove the straps from two watches, and obtain a single, extra-long strap. Attach it to both watches as shown in the drawing, and that's all there is to it. It works just as well, costs very little, and requires a minimum amount of work.

MATERIALS LIST

2 wristwatches (small) (see text)
1 pc. aluminum
1 rivet
-or-
1 extra long watch band
2 wristwatches (see text)

PROJECT 21

The Typewriter Watch

MANY PEOPLE ARE employed in one form of office work or another, where they like to have a constant check on the time but have no clock handy. Although nobody likes to be called a "clockwatcher," the work day does revolve around the clock. You time the start of the day, its finish, and the hours for lunch and coffee breaks as well. You ask your friends to call at a given time. Much office work centers on a typewriter, and that's the place to put a watch if you're going to have one.

There are two ways to make a typewriter watch, and which you choose depends on who owns the typewriter! If the machine belongs to the company, you can't easily mutilate it without getting into trouble. If it's your own typewriter, you can do with it as you will.

By far the easiest method is the one that is least

permanent. Visit your local fabric store and get some one-inch-wide Velcro ribbon. This is a two-part material, consisting of small hooks on one side, and small loops on the other. When the two are pressed together, the hooks engage the loops, and the ribbons adhere tightly. Pull them apart, and they disengage. This can be repeated as often as you like.

Cut a circle about the size of a dime from each piece, and use Pliobond cement to attach one piece to your typewriter. Glue the other piece to the back of your watch.

Now you can attach the watch to the typewriter, where it will stay in plain view all day. At night, before you go home, simply detach it, wind the watch, and put it safely in your locked desk drawer until morning.

For a more permanent installation, measure the face of the watch, and locate a place on the machine's lid where the watch movement will not interfere with the typewriter's works. Drill a small pilot hole in the lid at this point, and use a tapered reamer to enlarge the hole to three-eighths of an inch. Then use a round chassis punch of the correct diameter to cut the access hole for the watch.

Lift the typewriter's lid, insert the watch into the hole from the rear, and cement it into place with epoxy cement. To wind the watch each day, simply lift the lid. This should give you clear access to the winding stem.

MATERIALS LIST

1 pocket watch
1 small piece of Velcro
Pliobond cement

PROJECT 22

The Hippie Watch

ONE OF THE FIRST things that younger protesters protest is the regimentation of time. They don't dig it. They'll either *be* at a place whenever you get there, or expect you to just stick around until they make the scene. Don't rap about time, for it just doesn't groove.

Today's youth are much attached to symbols. They wear love beads with peace symbols dangling from them, or the symbol "ankh" for "life."

Here's a new symbol. It bespeaks a total contempt for time. It looks great on love beads and is easy to construct.

Start with a large, inexpensive pocket watch—the kind that you can get at any drugstore for about one dollar. The less expensive it is, the more loudly it ticks, and that's an asset.

Using a knife edge, remove the crystal from this

watch, and with a tweezers, lift off the hands—that's right, the hour hand, the minute hand, and the second hand. Now put the crystal back, attach the watch to a long chain of love beads, and wind the watch. Our own favorite hippie gets a big charge over winding the watch and consulting it occasionally.

As an added attraction, save the hands. We were asked to put the minute hand back in place. Why? Now when she's asked the time, our little hippie can say "It's half-past."

MATERIALS LIST

1 large, cheap pocket watch
1 strand of "love beads"

PROJECT 23

The Ring Watch

WRISTWATCHES ARE NOTHING NEW, but here's a new twist that makes a more-than-welcome gift for any lady. It's a ring that tells her the time and eliminates any need to wear a strap on her wrist.

Start with an inexpensive ladies watch movement which you can pick up at any novelty shop. Select a watch whose rear case snaps off instead of unscrewing.

Your next stop will be a jewelry shop, where you will buy a simple gold wedding band in the correct size for the recipient of this gift. Make sure there's some gold content in this ring.

Remove the wrist strap from the watch, and then remove the watch from its case by prying the back off. With a file, carefully take the four ears off the case, making it nicely rounded. Now you can either replate the case, or if it is gold-filled, polish away your file marks. Replace the works in the case, and set the case aside.

Using the file, clean up a surface on the back of the watch. If the back is lacquered, remove all this lacquer. Place the ring in a vise, and file one surface flat.

Now you are going to do some pretty fancy soldering. Obtain a length of silver solder from your local jewelry supply house, and then mix borax and water, which serves as a fluxing agent. Use a torch for this type of soldering. Paint the borax on the ring, only on the area you filed flat. Heat this surface with the torch until the ring is hot enough to melt the solder when it is touched to the ring. Allow a little to flow in place and wipe off the excess with a soft cloth. Repeat for the back of the watch, thereby tinning both surfaces.

Now place the back of the watch on a firebrick, place the ring on top of the back, and apply the heat again. Add a touch of the solder to add some metal, and allow the whole to cool off.

When it's cool, place the back of the watch in position on the watch case and snap it closed. There you have the ring watch, which needs only to be wound and worn.

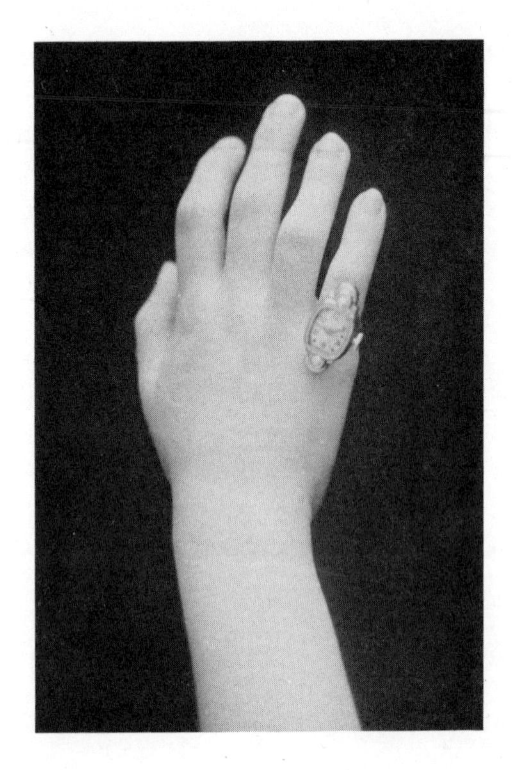

MATERIALS LIST

1 small ladies wristwatch
1 gold-filled wedding band
1 small length of silver solder
small quantity of borax mixed with water to heavy paste consistency

PROJECT 24

The Cuff Link Watch

TO MAKE THIS WATCH, you're going to have to do a bit of searching to locate two different kinds of things at two different places.

You will need either a square cuff link and a square watch or a round link and a round watch. To complicate matters, the cuff link must be just a trifle smaller than the watch, so do your shopping with a ruler in your pocket. But let's say that you finally find both a watch and a cuff link to suit your fancy.

You can't easily solder the watch to the link, for the support arm of a cuff link is usually soldered, and your attempt to solder the watch in place will melt the face of the link, causing more trouble.

Start by filing the band-support ears from the watch, and then dressing away the file marks. Now carefully rub the back of the watch case with sandpaper or emery cloth to remove any plating or lacquer that may have been used. Do the same for the face of the cuff link.

Mix up a small amount of epoxy cement, measuring with care, and blend it thoroughly. Apply a small amount to the cuff link, which should be supported properly in a vise. Orient the watch on the link so that with the cuff on your sleeve, you will be able to read the time. Press the watch down on the link, and move it a bit to spread the epoxy, quickly wiping up any excess that squeezes out. Be very careful not to get any of the epoxy cement on the watch's winding stem or into the winding-stem hole. If you do, you'll have a watch link that just won't tell time! Be careful not to get the epoxy on the crystal or on the side of the watch, where it might cement the watch's back permanently into place.

Now set the vise aside overnight and don't touch it until the epoxy has hardened. In the morning your new watch link will be ready to wear.

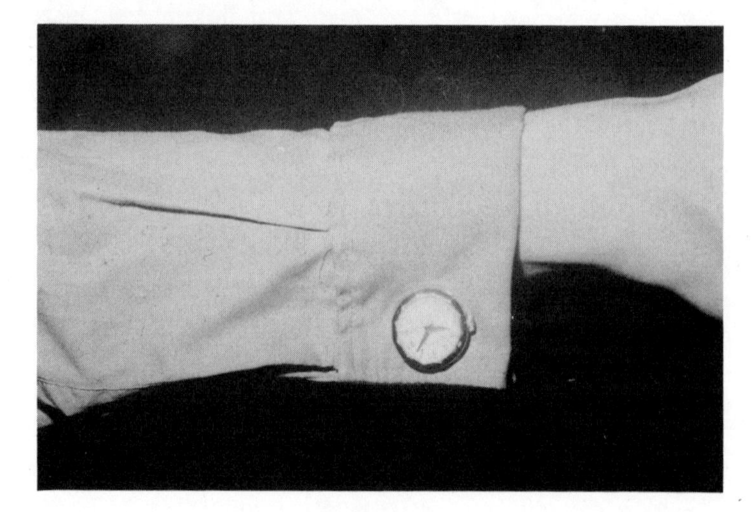

MATERIALS LIST

1 small watch
1 set of men's cuff links (see text)
small quantity of epoxy cement

PROJECT 25

The Dieter's Wristwatch

YOUR AUTHOR IS a pilot, and he's fat. As a pilot, he frequently wears a chronograph, which is actually an accurate stopwatch calibrated to give speed and which can be used to measure time as well as distance.

Your author is also constantly on a diet and has to count carbohydrate grams. He is allowed 60 grams of carbohydrate per day. And this is where the dieter's wristwatch is handy.

Use *any* stopwatch that's calibrated in 60-second increments. In the morning, set the second hand to the "12" and sit down to breakfast. When you finish the meal, start the watch, and stop it at the correct number of grams. If you've had a ten-gram breakfast, stop the second hand at the "2," which represents ten grams. Continue this during the day, and you'll have a running account of your daily gram intake. It's automatic, too.

If you are on a calorie diet, you will be allowed more than just 60 calories. The answer is a different sort of stopwatch—one designed exclusively as a stopwatch. This stopwatch has a second hand that counts to 60 and has a minute and hour hand that are also stoppable. This sort of watch is calibrated in tenths of a second and can count up to hundreds of thousands, more than enough for the most liberal of calorie dieters!

Dieters are ever on the alert for new means of keeping score, and this writer for one is delighted with his "automatic" counter. It beats the scales, dials, and other cardboard and plastic contrivances that we've seen till now.

You can locate suitable stopwatches for this purpose in the Lafayette Radio catalog.

MATERIALS LIST

1 stopwatch (see text)

PROJECT 26

The Wrist Hourglass

DO YOU HAVE a sense of humor? The author has gotten more laughs from the hip, and puzzled expressions from the un-hip, with this novelty wristwatch.

Visit a novelty shop in your area, or consult the many mail-order houses for one of the small, one-minute hourglass timers. The ideal size for our application is about one and one-half inches long. Depending on the materials, the cost should be under one dollar. Your local five-and-dime store will provide you with a toy wrist-watch that you won't mind taking apart.

Remove the crystal from the toy watch, and remove the works so that all that remains is the body and the strap, Cement the hourglass to the body of the watch, and allow the cement to dry overnight.

To use the watch, put it on your wrist, and put your real watch on the other wrist. When somebody asks you the time, lift your coat sleeve so they can see the "watch." Take a good guess at the actual time, keeping a straight face all through this. Another way to use the watch is to simply allow it to be seen—casually—by those around you.

MATERIALS LIST

1 Toy wristwatch (local five and dime store)
1 small hourglass
small tube of household cement

Where to Buy Parts

ANYBODY WHO sets out to buy clock or watch parts soon learns that this is a closed sort of deal. Manufacturers will sell parts only to recognized jewelry wholesalers, who, in turn, will sell only to recognized watchmakers.

The amateur hasn't got a prayer, for no jeweler or watchmaker will sell parts when he can make better profit by actually doing repairs!

Happily, the projects in this book do not require that you obtain watch parts, but rather, complete movements. The following firms will sell you not only the components you require, but the accessories as well. We have contacted them, and obtained their agreement to this end. However, before investing in new components, look around and see if you don't have a discarded clock or watch that still functions well. Often, crystals or cases become marred, and the movement is perfectly good.

Edmund Scientific Company
100 Edscorp Building
Barrington, N.J. 08007

Lafayette Radio Electronics
111 Jericho Turnpike
Syosset, L.I., N.Y. 11791

Lanshire Clock & Instrument Co.
c/o Empire
1295 Rice Street
St. Paul, Minn. 55117

Glossary

BX Cable: armor-sheathed electrical cable required by electrical codes in some cities for in-the-wall wiring

Cam: shaped wheel attached to a motor whose purpose is to convert rotary movement (the motor shaft) to lateral movement (the cam follower) usually used for operating switches or valves

Cam follower: a rod, usually equipped with a frictionless roller, that follows the curve of the cam

Clock fit-up: a complete clockwork consisting of mechanism, face, hands and crystal

Clock radio: a radio receiver and clock, connected so that the clock always operates, and can be used to turn the radio on or off at the preset times

Candle clock: used by ancients; consists of a candle of fixed diameter that would burn at predetermined rate; length of candle is marked off in hours

Cordless clock: an electric clock that operates from a battery source, as opposed to one that operates from a plug-in electrical source

Crystal: the glass or plastic face or bezel of a clock

Crown: the winding stem of a wristwatch

Electric clock: clock that operates from an electric source instead of being hand wound

Electronic clock: clock mechanism that operates from an electronic oscillator which maintains time and accuracy

Electronic flash: high-intensity light source used in photography

Escapement: geared or toothed wheel that operates with a pawl or rocker arm so the stored energy in a spring is released in a controlled fashion

Ferrotyping: process used in photographic enlargements that imparts a shiny surface

Fire brick: type of brick used in fireplaces and cooking pits that is resistant to heat

Flexible gooseneck: spiral-formed, hollow rod that flexes, used in making lamps

Gem box: electric utility box for in-wall mounting of electrical components

Gusset: support for corners and edges of boxes; usually triangular in cross-section

Hourglass: sand-filled glass container with a stricture in the middle; measures time it takes for sand to empty from top to bottom

HOUR HAND: hand on a clock (usually the smallest hand) for recording the passage of hourly increments

LAMP HARP: bowed metal section of lamp used for supporting the shade

LUMINOUS DIAL: watch face treated with special paint that glows in the dark, making hands and numbers readable

MINUTE HAND: middle-sized hand on clock used for recording passage of minutes. Makes one full rotation each hour

OIL LAMP CLOCK: ancient timepiece that burned oil at a regular rate. Glass container for oil is calibrated in hours

OPAQUE PROJECTOR: device for projecting opaque images (as opposed to slide or transparency projector)

PAWL: rocker arm of escapement (see Escapement)

PENDULUM: swinging arm used to regulate speed of clock

PENDANT SWITCH: electric switch designed to suspend from a cable end

PAINT FILLER STICK: hard enamel paint in stick form that is applied by rubbing over scratch marks or engravings

PILOT LAMP: small lamp used in electronics to indicate that a circuit is "on"

POP RIVETOOL: hand tool for installing rivets by pulling a wire which turns the rivet tail to make the joint

PRINTED CIRCUIT BOARD: copper laminated to a phenolic base, usually etched to form conductor strips and component mounting areas

ROMEX CABLE: insulated electrical cable, approved for in-wall wiring in some cities

RUB-OFF TYPE: printed type in various sizes and faces, mounted on waxed sheet, applied by rubbing on surface

SCRIBER: pointed tool used for marking wood or metal

SECOND HAND: small hand of clock that makes one full sweep each minute

SOLDERING: process of using low melting point metal to join other metals by applying heated tool

SPEEDBALL PEN: special marking pen with broad nib for making heavy lines

STOPWATCH: watch with second hand that can be started and stopped by pressing small buttons

SUN DIAL: ancient clock that used the position of the sun and the shadow it cast on a special plate to record the time

SWEEP SECOND HAND: second hand of clock or watch that is mounted concentrically on the hub or armature; is the longest hand on unit, sweeps the entire face with each rotation

SWITCH: electrical device used to interrupt flow of current

TIME: method of breaking up periods of day, hour or minute in such a manner as to be recordable

TINNING: process of applying small amount of solder to a surface making it more easily "wettable" by additional solder, prior to bonding to other metals

TRANSFORMER: electrical device to raise or lower amount of voltage; a step-down transformer lowers the input voltage

VELCRO RIBBON: two-piece cloth ribbon, with small hooks on one side, small loops on the other. Pressing the two pieces together joins them in a firm bond until they are removed by pulling apart

WATER CLOCK: calibrated glass with stricture at bottom through which water drips at a measured rate

ZIP CORD: two-conductor electrical cord